The Miracle Molecule

The Biological Importance of Water

Why I Wrote This Book

Without it, life would not exist on this planet. It is essential to all living things. It is an extraordinary molecule with extraordinary properties. This is the only planet in the Solar System where its liquid form has been found. All of us humans are made up of around two thirds of it...

I am talking of course about: water - the miracle molecule.

What I aim to do in this book is explain the biological importance of water in terms of water's unique chemical properties.

Book Contents

Chapter One - The Chemistry of Water

Before you get to grips with the biological importance of water it would be really helpful to understand its chemistry.
So here are some relevant facts about water - it really is an amazing little molecule!

1. Water has the chemical formula H_2O.

- This means that a water molecule has 2 hydrogen

atoms and one oxygen atom in it.

- Water is therefore a relatively small molecule, particularly as hydrogen is the smallest atom and oxygen the sixth smallest.
- Water is also a very light and low density molecule – as it weighs only 18 atomic mass units (as a comparison, carbon dioxide, CO_2, weighs 44 atomic mass units).

2. Water is a simple covalent molecule.

- This means that the oxygen and hydrogen atoms share two electrons (they each contribute one electron to the bond) which help to form a strong bond between the oxygen and hydrogen atoms.

- As this bond is quite strong water molecules do not break apart easily and cannot usually be broken apart by heating.
As this bond is quite strong water molecules do not break apart easily and cannot usually be broken apart by heating.
However, water does take part in some chemical reactions (including biological ones) where one of the bonds between hydrogen and oxygen break.

3. Water has hydrogen bonding.

- Water has very strong forces

of attraction to other water molecules – these inter-molecular bonds are called: 'hydrogen bonds'. The strong attraction is due to the regions of opposite charge on the water molecule and the small size of hydrogen atoms. The partly positively charged (hydrogen atoms) and partly negatively charged (oxygen atom) regions attract between different water molecules.

- The strong hydrogen bonds between water molecules mean that water has a very high melting point (0 °C) and boiling point (100 °C) for such a small molecule. As the average temperature on Earth is about 15 °C this means that there is plenty of liquid water on Earth, one of the main conditions necessary for life. (At the North and South ends of the Earth it is so cold that water is almost permanently in its solid state – ice; while in the deserts water only exists as a gas – water vapour).

4. Water is a polar molecule.

- This has nothing to do with the Earth's North and South Poles at all. What it means is that water molecules have slightly positively charged and slightly negatively charged regions (-the negative is around the oxygen atom and the positive around the hydrogen atoms).

- The fact that water is a polar molecule is very important. It means that water can dissolve almost all charged particles by surrounding them and taking them into solution. It is thanks to this fact that water is the universal biological solvent, forming three-quarters of the cytoplasm of all cells.

- The fact that water is a polar molecule means that water can form very strong hydrogen bonds with other water molecules (see above).

5. Water is a 'V' shaped molecule.

- The water molecule is a 'V' shape, with oxygen being at the sharp end of the 'V' and the hydrogen atoms at the ends. This shape, along with the strong hydrogen bonds that exist between water molecules mean that when water forms a solid, it can form an open structure with many 'V' shaped water molecules, tessellating and sticking to others. This means that solid water (ice) has a low density, even lower than liquid water – resulting in ice floating on water. Water is the only known substance in the universe whose solid floats on its liquid form.

Chapter Two – How Are The Properties of Water Biologically Important?

1. Water is an excellent solvent.

Fact → Water can dissolve almost all molecules found in living things: this includes: amino acids, globular proteins (e.g. enzymes and blood proteins), sugars (mono and disaccharides) as well as mineral ions such as iron and sodium.

Water is an excellent solvent because it can dissolve almost any biological molecule that has a positive or negative charge - such as sodium and chloride ions and amino acids - and anything that has polar chemical groups (which have small positive and negative charged regions) such as proteins and sugars.

(The only biological molecules water cannot dissolve fully are fats or fatty acids and these usually form a suspension in water.)

Biological Importance

- Almost all living cells (and

therefore living organisms) are made up of around 75% water because water is the universal biological solvent in which the chemical reactions that drive cell metabolism, such as respiration, take place. It is no exaggeration to say that liquid water makes life on Earth possible.

2. Water is the only substance which has a solid state (ice) which floats on its liquid state (liquid water)

Fact → Ice is less dense than liquid water and this means that ice floats on water, something that we take for granted, but it is in fact astonishing that a solid substance can float on its own liquid. Water is literally the only known substance in the universe of which this is known to be true.

Biological Importance

- For lake, river and sea dwellers, ice floating on water means that the lake, river or sea water is unlikely to freeze right down to the bottom, the reason being that ice floats and forms an insulating layer on top of the water that helps to prevent further freezing below it. If a lake, river or sea froze to the bottom most of the living things within it would die due to lack of oxygen and food, so the fact that ice helps prevent this is very important.

3. Water heats up and cools down very slowly.

Fact → Water has a very high specific heat capacity. This means that it takes lots of energy to heat water up; and that water needs to lose a lot of energy to cool down.

Biological Importance

- Many organisms live in water. The fact that it requires a lot of energy change to heat or cool water means that water is a fairly constant

environment in terms of temperature. It also means that land based organisms will heat up and cool down much slower (as almost all organisms are made up of around 75% water).
This fairly constant temperature is very important to living things, as it means the enzymes that catalyse and make possible the chemical reactions (such as respiration) that make life possible, can function fairly well most of the time.

4. Water is very sticky to itself

Fact → Water molecules are very sticky to other water molecules.

This is termed 'cohesion'.
It is a consequence of the strong
type of attraction that happens
between water molecules, called
'hydrogen bonding'.

Biological Importance

• The Transpiration Stream
The transpiration stream is the
continual flow of water through
a plant from its roots to its
leaves through the ultra-thin,
hollow tubes of the xylem
tissue.
The water is pulled up the
xylem by the creation of
negative pressure in the leaves
by the loss of water through
transpiration (this force is
rather like that created when

we suck water up through a straw). But this would not work if water wasn't cohesive (sticky to itself). If water was not cohesive the continuous column of water would break apart and the water could not be drawn up the stem.

• Surface Tension

Surface tension is caused by the strong cohesion between water molecules. Some insects (such as pond skaters) ride upon the surface tension created by water.

5. It takes a lot of energy to boil water

Fact → Water has a high specific heat of evaporation. This is due to the strong forces of attraction – hydrogen bonding – between water molecules, resulting in a lot of vibration energy being needed (and therefore heat energy needed) to break them out from the liquid phase to the gas phase.

Biological Importance

• Sweating and Panting
Because water has a high heat of evaporation, sweating takes a lot of heat energy away from the skin - which is great because that is exactly what sweating is for; it is an adaptation found in some mammals (e.g. humans, horses and pigs) that enables cooling of the body and the blood.

The sweat is produced by tiny sweat glands in our skin that produce salty water, under the instruction of the autonomic nervous system. The salty water is secreted onto the surface of the skin when our bodies core (internal) temperature rises e.g. during hot weather or exercise. The evaporation of this water cools the skin and blood, thereby cooling the core when the blood returns to it.

This effective cooling mechanism keeps the body core temperature at around 37 °C in humans.

However, dogs are mammals but cannot sweat as they are covered in hair – to cool down they pant instead, cooling the blood by evaporating water from their mouth and tongue.

6. Water Is Involved In Many Biochemical Reactions

Fact → Water is involved in both hydrolysis and condensation reactions.
Hydrolysis reactions take place during digestion.
All digestive reactions are hydrolysis reactions.
Hydrolysis means splitting water, and water is indeed split up during these reactions.
Condensation reactions are the opposite of hydrolysis reactions.
Water is created during these reactions.

Biological Importance

- Hydrolysis reactions split up biological polymer molecules, such as during chemical digestion. For example the reaction: starch + water → maltose; which takes place in the mouth and small intestine is a typical hydrolysis reaction, water splits up the glycosidic bond in starch (helped by the action of the active site of the enzyme amylase) to leave maltose sugar, a disaccharide.
- Condensation reactions tend to happen in anabolism, which is the building up of biological polymers such as proteins, polysaccharides and DNA molecule. For example the joining of two amino

acids is a condensation reaction: amino acid + amino acid → dipeptide + water; notice that water is produced in this reaction as the peptide bond is formed between the two amino acids. As amino acids are joined in this way; and polypeptide chain polymer is formed which forms the basis of a protein molecule.

Chapter Three - Humans and the Homeostasis of Water

Why Does The Body Control Its Water Level?

As with the great majority of living things, humans are made up of about 75% water.
This is because our cells and blood are both around 75% water.
As explained earlier, this is due to water being an excellent solvent for biological molecules.

Our bodies are equipped with a mechanism that controls the amount of water they contain.
If we have too much water in our bodies, our cells will swell up and eventually burst, also the blood volume would increase greatly, causing dangerously high blood pressure.
If we have too little water in our bodies, our cells will shrivel up and cease to function and in addition our blood volume would decrease greatly, causing dangerously low blood pressure.

So both having too much and too little water in our bodies can be dangerous and unhealthy, so we have a hormonal mechanism that works between the pituitary gland and the kidney, to try and balance the amount of water in our bodies.

Let's start by looking at how the body might lose or gain water, causing the amount of water in the blood to start to either decrease or increase respectively.

How Does The Body Lose Water?

The body loses water in three ways.

1. Sweating. Water evaporates off the body, after being produced by tiny sweat glands in the skin. This is the body's cooling mechanism, preventing the

body's core from getting too hot. We tend to sweat more in high environmental temperatures and during exercise, although some sweat is lost every day. Anything between one liter (litre – UK English) and ten liters is lost every day through sweating.

2.   Urination. Water leaves the body in the urine when we urinate. Urine is a water-based solution of urea (a toxic waste product of protein metabolism), ions (such as sodium and hydrogen) and non-self chemicals such as medications that might be present in the blood. The

body's homeostatic mechanism (explained later in this chapter) will cause the increase or decrease of water loss in the urine, depending on the level of water in the body and blood at that time. An average of approximately between 1 and 2 liters ('litres' in UK English) of urine is produced every day.

3. Breathing. You may notice that when you breathe out in cold weather or onto a cold glass pane, you can see 'mist' that is the condensed water vapor that is present in your breath. This is a result of the lungs having a damp inner surface and the water will therefore leave by

evaporation when we breathe out. We would tend to lose more water in this way in hot and dry air conditions, as this type of air would tend to cause more evaporation from the inner lung surface.

4. Defecation. Feces – which is the substance that we get rid off from the anus when we go to the bathroom - contains some water. The amount of water it contains will vary, but if someone has diarrhea it can be a quite large amount of water lost from the body in this way.

How Do Our Bodies Gain Water?

The obvious answer is through drinking and eating (quite a lot of food contains considerable amounts of water, from the obvious example of soup, to juicy fruits and some vegetables, such as cucumbers).

However there is a second, hidden way in which we gain water. This is called metabolic water. When we respire, glucose is converted in aerobic conditions to carbon dioxide and water, and this water is what is termed 'metabolic water'.

On average we need to consume at least two liters of water, in one form or the other (in pure water, milk, soups, fruits etc) a day in order to replace that lost in sweating and urinating.

What is Homeostasis?

Homeostasis is the principle of the body trying to keep levels of things within an acceptable range that is healthy and not damaging. The level of sugar and salt are both controlled in this way, as is the level of water.

Homeostasis does not mean keeping levels exactly constant all the time. There are fluctuations in the levels but the homeostatic mechanism causes them to only fluctuate to a lesser degree, keeping the levels with acceptable limits.

What is the mechanism by which the level of water is controlled in the body?

A hormonal mechanism works using the hormone ADH and the kidney as a target organ.

How Does The Mechanism Work?

It works by the principle of negative feedback.
In other words, if the amount of water in the blood rises too high, the mechanism acts to bring it down again. On the other hand if the amount of water in the blood falls too low, the mechanism

acts to bring it back up again, or at least to stop it falling further.

So to control the amount of water in the blood the body uses the principle of negative feedback.

If the level of water in the blood starts to fall, this is detected by a part of the brain called the hypothalamus.
The hypothalamus then communicates with the pituitary gland (which it is right next to) causing it to secrete the hormone ADH (Anti-Diuretic Hormone) into the blood.
ADH flows in the blood to the target organ, the kidney, where it acts on the second (distal)

convoluted tubule and the collecting duct, causing them to take out more water from the urine and return it to the blood. This helps to reduce water loss from the blood at the kidney, acting therefore to oppose the lower water level in the blood. The result (you might notice this yourself) would be less, more concentrated urine, and fewer trips to the bathroom (toilet – UK English).

On the other hand if the level of water in the blood starts to rise, this is detected by the hypothalamus, which then instructs the pituitary gland to stop releasing ADH into the blood. The kidney then removes less water from the blood, so more water ends up in the urine. This acts therefore to oppose the increase in the water level in the blood by removing water from the blood in the urine. The result – you might find yourself going to the bathroom (toilet – UK English) more often and producing a greater volume of dilute urine.

www.ingramcontent.com/pod-product-compliance
Lightning Source LLC
Chambersburg PA
CBHW061452180526
45170CB00004B/1671